**TO FLEDGLING FLOWER LOVERS,
TENDING THEIR FIRST OF MANY BLOOMS.**

We acknowledge the First Nations people as the Traditional Owners
of the land, air and waterways. Sovereignty was never ceded. This was written
on lands of the Bunurong Boon Wurrung people of the Kulin Nation and
we pay our respect to their Elders past and present, and Elders of all
First Nations on the lands on which we live, work and garden.

THE

SUPER BLOOM HANDBOOK

BY JAC SEMMLER
WITH PHOTOGRAPHY BY SARAH PANNELL

CONTENTS

INTRODUCTION

Hello fledging flower lover! Have you always wanted flowers at your fingertips but were not sure how to get started? Or even what to do to grow beauty? *The Super Bloom Handbook* is for you, your flowery friend to get you through those modern-day realities that keep us from the garden.

Here are flowers for any budget or for even the smallest, most temporary space you might live in. Here are flowers that are abundant and resilient, that will grow even if you garden in tough conditions. Here are self-sufficient beauties to grow even if you snatch gardening in small moments of time, so your garden can overflow with delight even with overflowing calendars. With this book, you will discover the beauty that nature can offer, and the joy of tending plants and holding flowers in your hands.

Embrace the blooms. There is so much beauty and wonder to find in growing flowers. You *can* grow these super blooms.

OPPOSITE • Chamomile and Borage (see p. 36), two flower friends both loved by bees.

HOW TO START

CHAPTER 1

This book is a call to garden in any way you can. You do not need permission or expertise – we all start somewhere. Gardening is a process, something that builds upon every season through time and tending.

HOW DO YOU START ON THIS JOURNEY?

1. Go on a mission to find the plants you love and that you think will grow well with your care in the garden. Things flourish when they are tended to.
It is easy to feel overwhelmed, but simply starting is the most important thing.

2. Start small. Choose a location and fill it with a plant or plants that appeal to you. Lavish it with care and watch it grow. If it doesn't thrive, be curious. Ask, why did this plant die? Where does it need to be? What can I do to make it happier? Should I grow a different flower instead?

3. Experiment with colour, shape and texture. No combination of these elements is wrong. The most important thing is pleasure in what we grow and wonder in what we find. Gardening is an expression of you, not a strict set of rules to be followed.

4. If something you like isn't growing well, find plants that display the same shape or colour you were initially drawn to but are better suited to your climate. The natural parameters of location and climate will always guide gardening. These conditions are opportunities to be curious and creative.

This handbook includes over forty plants with brilliant blooms for various climates and conditions. It is not a definitive list, simply a starting point of new and old flower favourites to consider and grow that are generally easy and accessible in busy modern lives.

THE CHAPTERS COVER: Flowers that can provide maximum blooms on a budget; full flower power for small spaces; resilient flowers for enduring beauty in tougher conditions; and abundant and self-sufficient flowers you can enjoy if time is tight.

Each plant profile is structured with straightforward information to help make a start.

✽ **WHY** — Why you must grow this flower in your garden.

✪ **WHAT** — What to do to start. Grow it from seed, a punnet of seedlings, or a cutting.

✦ **WHEN** — When in the year to start planting or propagating.

☀ **WHERE** — Where to plant and cultivate these flowers, the conditions and situations they like best.

➤ **HOW** — How to care and tend for these flowers.

A NOTE ON PLANT NAMES
Plant Profiles feature both the common plant name and its botanical name, which is set in italics. Sometimes these names are the same but often they are different. For example, the botanical name for Sweet Pea is *Lathyrus odoratus*, while *Pelargonium* is both the common and the botanical name. Botanical names are used to identify the genus and species of a plant and can also include a variety or cultivar name, for example, *Salvia nemorosa* 'Caradonna'.

GENUS	SPECIES	'CULTIVAR' OR VARIETY (VAR.)
Salvia	*nemorosa*	'Caradonna'

PREVIOUS • The delight of a Poppy's papery petals (see p. 22).

OPPOSITE • An elegant Foxglove flower spike (see p. 78).

One of the most exciting elements of a garden is the relationship plants have with each other. Flowers are beautiful in their individuality, but bringing two or more different plants together creates a special dynamic. As plants grow, bloom and decline in a garden, they create a series of beautiful sights.

When it comes to growing flowers together, look to their ideal growing conditions as a guide to their compatibility as flower friends. Choose plants that need the same levels of light and water to increase the likelihood of them thriving side-by-side as neighbours. You might choose two plants that both enjoy shade and fertile soils improved with organic matter, or plants that both need full sun and seasonal watering through summer.

Choose plants to produce a homegrown super bloom, with a riot of flowers blooming at the same time, or extend the wonder by planting flowers that bloom one after the other for a year-round calendar of flowers.

As you discover flower friends you love, it is easy to keep adding plants to create more diversity and seasonal wonder but it is important to consider the spacing of plants so they can grow into their lovely forms without swamping their neighbours.

Many profiles in this book include suggested flower friends. There are endless possible combinations to create beauty in your garden.

A NOTE ON LOCATION AND SITUATION

We are all gardening in different situations and locations. More and more of us live in small, temporary places or come home to a concrete backyard with no direct light. However, these restrictions can lead us to extraordinary and unexpected discoveries in new ways of gardening and new plants to grow. This book is a creative call to arms, a rallying cry to cultivate beauty with what you have.

Advice is general, accessible, applicable across different climatic regions and based on the season (rather than month). For all the flower lovers tending flowers in different climates, there is a broad selection of flowers to experiment with in your heartland.

OPPOSITE • Flower friends in action: Umbellifer Wild Parsnip (see p. 88) with red Kangaroo Paw (see p. 146) and *Agastache* 'Sweet Lili' (see p. 100).

MAXIMUM BLOOMS ON A BUDGET

Take a clever approach for a garden brimming with
big-hearted beauty on any budget. These flowers can easily grow
from a simple packet of seed or form roots from a cut stem.

A GARDEN CHAMPION

✷ WHY
Trailing stems of Nasturtiums ramble across a garden, creating a lush, verdant scene that can envelop any fence or frame.

◑ WHAT
Grow Nasturtiums from seeds or seedlings. Punnets of plants are occasionally available in mixed colours or single varieties from quality nurseries.

⟁ WHEN
Sow seed direct in spring to early autumn if you live in a frost-free area. The same advice holds for when to plant seedlings from punnets into the garden or planters.

☀ WHERE
Nasturtiums enjoy full sun and tolerate part-shade positions. They cope with many soil types and conditions and often flower most profusely in poorer soils.

↘ HOW
Nasturtiums germinate from seed into the cutest seedling with sweet lily pad leaves emerging. They are prolific self-seeders, so once you grow them through a few seasons you will find new seedlings volunteering their time.

❀ VARIETIES
There is a huge range of Nasturtium varieties with many flower colours to revel in and enjoy.

Favourite varieties of Nasturtiums

'RED EMPEROR'
Brick-red flowers that fade like fabric with streaky details on the petals.

'EMPRESS OF INDIA'
Regal, deep-jade leaves; dark like a green-blue sea.

'SALMON GLEAM'
Ruffled peach flowers and a darker throat than any other variety or cultivar.

'ORCHID CREAM'
Cream-coloured flowers with delicate, red-veined petals. Compact and dramatically beautiful.

TIP • These beauties are a wonderful companion plant in the vegie garden as they attract a glorious array of pollinators. The flowers, leaves and even the green seed pods are edible.

PELARGONIUM

A LONG-FLOWERING WONDER

✳ WHY Resilient in tough, dry summer conditions, the infinite varieties of *Pelargonium* fill pots and gardens with striking flowers and spectacular foliage.

✿ WHAT *Pelargoniums* are one of the easiest plants to grow from cuttings **(see p. 154)**. Small cuttings are so easy to share between friends or – if you're feeling cheeky – they can be discreetly plucked from neighbourhood gardens, street plantings and abandoned land.

⚘ WHEN *Pelargonium* cuttings form roots with ease in the warmer parts of the year in spring and early summer. Plant in autumn or spring in temperate climates or in spring in cooler climates if there is extreme cold.

☀ WHERE Full sun in a free-draining position. Grow as perennials in hot and temperate gardens or as annuals in cooler climates.

☙ HOW *Pelargoniums* are sensitive to frost and do not like wet feet in the cool months. Fertilise with organic slow-release pellets. Deadhead regularly to encourage ongoing flowering.

✿✿ VARIETIES & SPECIES
Pelargoniums with unique features

PELARGONIUM GRAVEOLENS
Best known as the Rose Geranium with indulgent fragrant foliage – a treasured favourite.

PELARGONIUM 'CHOC MINT'
A delicious favourite for its sizable, soft, velvet foliage with deep chocolate centres.

PELARGONIUM SIDOIDES
Scalloped leaves and delicate fine flowers sit on fine stems about the mound of foliage.

♥ FLOWER FRIENDS

SALVIA
AGASTACHE
STATICE
SEDUM
ERYNGIUM

BEARDED IRIS
COSMOS
TEUCRIUM
VERBENA
CALIFORNIAN POPPY

THIS PAGE • The luminous foliage of *Pelargonium* 'Choc Mint'.

POPPY

PAPAVER RHOEAS, PAPAVER NUDICAULE, PAPAVER ORIENTALE

A FLOWER THAT WILL STEAL YOUR HEART

WHY

The alluring quality of the delicate paper-like petals of annual poppies is just too hard to resist. When it comes to Poppy, don't hold back! A mass of poppies with a multitude of flowers blooming together is an utter joy.

WHAT

Annual Shirley and Icelandic Poppies are abundant germinators from seeds but can also be found in seedling punnets at nurseries. Opium or Peony Poppy seeds are not commercially available in some countries, being a regulated crop, but their plump flowers and ornate seed heads often appear in old gardens.

WHEN

Shirley and Icelandic Poppies can be sown by seed directly into the soil from autumn to spring.

WHERE

Annual Poppies enjoy full sun and free-draining soil and will freely self-seed, encouraging the next year's paper petal blooms.

HOW

Sow directly onto finely raked soil, keep damp until germination. Sow in trays or punnets and plant seedlings densely spaced.

VARIETIES

Annual poppies for infatuations

***PAPAVER* 'LADY BIRD'**
A brilliant and bold variety with a charming black splotch for every scarlet petal, just like its namesake.

***PAPAVER* 'PANDORA'**
A luscious variety with a spectrum of plum and scarlet tones. Delicious.

***PAPAVER* 'AMAZING GREY'**
Though not as robust as other varieties, 'Amazing Grey' is truly wondrous with its pale grey shades of petals. A fun curiosity to tend.

FLOWER FRIENDS

CORNFLOWER
SCABIOSA
NICOTIANA
BORAGE
FOXGLOVE
BEARDED IRIS
ROSE
GEUM
SALVIA
PELARGONIUM
PENSTEMON
BILLY BUTTON
ORLAYA
SEDUM
HOLLYHOCK

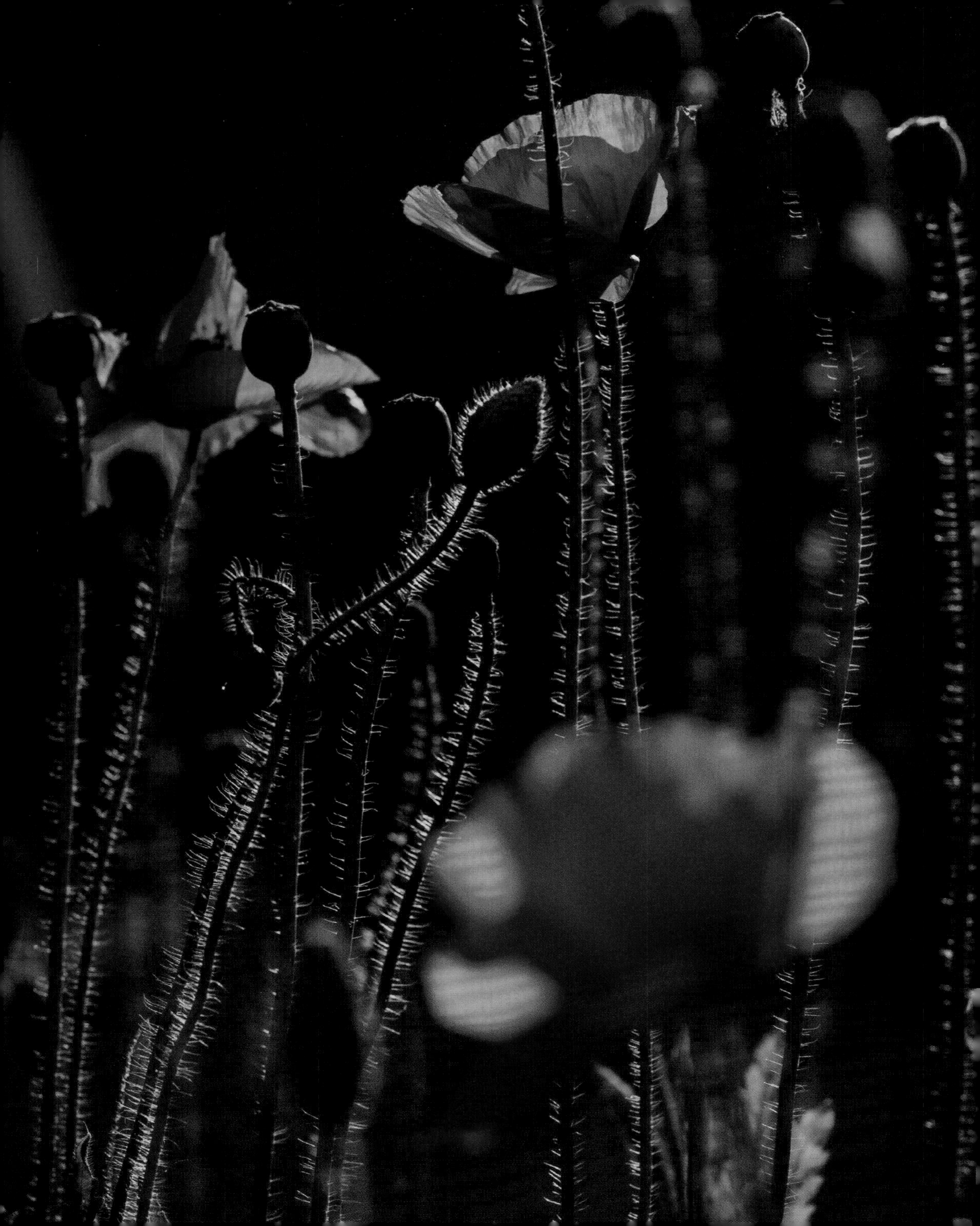

TIP • I have recently had great success from a surprising method where seeds are frozen in iceblocks before scattering into the garden. Fill an ice tray halfway and freeze. Sprinkle Poppy seeds in each cell before topping up with water and freezing again. Then scatter these iceblocks into the garden and water regularly until the seedlings start to germinate. Plants are surprising.

SUNFLOWERS

HELIANTHUS ANNUUS

ICONIC TOWERING BLOOMS OF SUNSHINE AND HAPPINESS

WHY Queens of a flower garden.

WHAT Sunflowers can be readily grown from seeds. It is amazing just how tall and fast they can shoot up in the warmer seasons, from a seed to a prolific, tall bloom.

WHEN Sunflowers enjoy germinating and growing in the warmer seasons so sow and plant seedings from late spring to early summer.

WHERE Sunflowers need sun, and lots of it in a sheltered position, as they can be top heavy. Plant closely for a self-supporting group or stake.

HOW Prepare well-draining soil with lots of organic matter. Scatter slow-release, organic fertiliser around the bases of seedlings (as per application instructions).

VARIETIES Sunflower varieties are either single-stemmed or multi-stemmed. The single-stem varieties only produce one sizeable, prominent flower for the season. Multi-stem varieties have smaller abundant flowers in a cluster and are often a bit shorter.

FLOWER FRIENDS

- **FENNEL**
- ***RUDBECKIA***
- ***NICOTIANA***
- ***COSMOS***
- ***SCABIOSA***
- ***CANNA***
- ***KNIPHOFIA***
- **UMBELLIFERS**
- ***DAHLIA***

TIP • Multi-headed Sunflower varieties can be tip pruned **(see p. 158)**. Harvesting and deadheading spent flowers will encourage more blooms if there is time in the season. Don't tip prune single-headed varieties – they only bear one bloom before bowing out. Early staking can help support large varieties. You can source your own seeds after your first Sunflowers with a nifty bit of seed saving **(see p. 150)**.

AN INTOXICATING BLUE TO BEHOLD

✳	**WHY**	Annual Cornflowers are the brightest cobalt blues, a completely enthralling flower colour you can easily grow in the garden from a packet of seeds.
◔	**WHAT**	Cornflowers can be germinated from seeds or sourced as seedlings in punnets. Save seeds every year for a continual supply of these bold blue beauties **(see p. 150)**.
🌱	**WHEN**	Sow seeds directly into soil or into punnets for transplanting as seedlings. Sow or transplant in the autumn and again in spring.
☀	**WHERE**	Full sun and a well-drained position is best for these flowers.
🗤	**HOW**	Adding organic matter to the soil or using quality potting mix if growing in planters will get them off to a great start. Keep any seeds well-watered until germination and protect tender tasty seedlings from pests. Regular deadheading of the spent flowers **(see p. 158)** will extend the flowering season.

❤	**FLOWER FRIENDS**	POPPY EVERLASTING *SCABIOSA* FOXGLOVE *NICOTIANA*	*SILENE* *ORLAYA* BILLY BUTTON NASTURTIUM WINGED EVERLASTING

TIP • Although known for their rich and enthralling blue, Cornflowers also come in intense shades of pink, burgundy and mauve. The Cornflower keeps its strong colour in its edible petal after drying, preserving their beauty for use in creative and culinary fun.

AN EXPLOSION OF HAPPINESS

✿	**WHY**	It is hard to not be a fan of the *Calendula*. Their open disc-shaped blooms are like radiators, emitting a warmth to make a whole garden just that bit more cosy.
◔	**WHAT**	There is so much thrifty value in a packet of *Calendula* seeds. Or grow from seedling punnets and young plants.
⚑	**WHEN**	Sow direct in early spring or in autumn. A cool, hardy annual that will glow through the cooler months.
☀	**WHERE**	Full sun.
✍	**HOW**	Tolerant of most soil types and excellent companion plants, as their open flowers attract beneficial insects. Remove spent flowerheads to promote further flowers.
✿	**VARIETIES**	*Calendula* vary in colour and the density of their petals. Some varieties are double-petalled and packed tight to their centre. In contrast, others have an open disc and bloom in shades of rich orange through to burnished apricot and yellow.

♥ **FLOWER FRIENDS**

COSMOS	ECHINACEA	POPPY	CALIFORNIAN
SNAPDRAGON	CORNFLOWER	BEARDED IRIS	POPPY
ERYNGIUM	EUPHORBIA	FOXGLOVE	RUDBECKIA
BORAGE	STATICE	DIANTHUS	

TIP • Harvest flowers just as they start to open their big, vibrant
blooms if you want them to last longer in vases. Leave some
flowers to develop fully and set seed to collect **(see p. 150)** or leave
dried blooms to self-seed in the garden.

DRUMSTICK

A FLOATING SUN IN FLOWER FORM

❋ WHY
Billy Button is a common name stretching across several Australian wildflowers. Also known as Drumstick (*Pycnosorus globosus*), it is so much more than a button with an incredible yellow sphere for a flower held aloft.

✐ WHAT
Drumsticks can be grown from seed but also as young plants sourced at small pot sizes (tubes) from nurseries specialising in cut flowers and Australian plants.

🌱 WHEN
Sow seeds in spring as the germination is more even and faster with warmer soil conditions. Young plants can be planted at any season outside the extremes of summer.

☀ WHERE
Full sun in free-draining soil. The more of these glorious sun-like flowers the better. Grow in dense clusters by planting in closely spaced groups.

✎ HOW
Low-phosphorus fertiliser is best for feeding. Water well during summer to increase the alluring flowers.

♥ FLOWER FRIENDS

ERYNGIUM	*ALOE*
COSMOS	*SEDUM*
SCABIOSA	*COREOPSIS*
VALERIAN	*DIANTHUS*
LYCHNIS	*CORNFLOWER*
ERIGERON	

TIP • The long thin stems of the Drumstick leading up to the golden orb create a floating impression. They are excellent flowers to harvest and dry by hanging upside down in a cool, dry environment.

A BEST FRIEND TO BEES

✿ WHY   As well as hosting a loud hum of bees on a sunshine day, there is a bit of magic about the plant itself. The light catches the soft fuzz of the foliage, illuminating the arching stems that hold aloft the delicate, starry flower.

✐ WHAT   Grow Borage from seed and it will self-seed.

⚐ WHEN   Sow seeds in autumn and spring as an annual flower and it will tolerate the coldest season. These are also great times to find and plant any young seedings or transfer the self-seeded volunteers in your garden into ideal locations.

☀ WHERE   Full to partial sun position in free-draining soil. Allow plenty of room for the large branching arms of Borage to grow. Consider a sheltered position as stems become brittle at the base.

☜ HOW   Sow seeds directly, thinning seedlings to allow plenty of room for this large plant to grow. As the plant grows heavy with blooms, the stems can collapse at the climax of the flowering season – a victim of its success! You can prevent this by using stakes.

♥ FLOWER FRIENDS

CALENDULA	*CANNA*
CHAMOMILE	MARIGOLD
FOXGLOVE	*SALVIA*
POPPY	SUNFLOWER

TIP • The dainty and delicate flowers, held upside down by hairy bracts and stems, are edible. Carefully pluck the flowers from the bract to collect the tiny blue flowers with star-shaped petals.

A GARDEN'S BRIGHTEST BLOOM

❋ **WHY**
A cheery, happy flower, the Marigold is a bright staple for the garden.

✿ **WHAT**
Marigold can be grown prolifically from seed and is also widely available as seedlings in the warmer months.

⚹ **WHEN**
Sow seeds as the season warms in springtime. Punnets of seed benefit from some extra warmth. Spring is also the time to plant seedlings directly into the garden or to fill out planters for more riotous colour.

☀ **WHERE**
Plant in full sun with fertile, well-draining soil. Enrich the soil by incorporating organic matter before planting or, if in planters, quality potting mix is ideal.

🢖 **HOW**
After sowing seeds in punnets in spring, transplant them to the garden or a pot when the temperatures are warmer and the risk of frost has passed in cooler parts. Regular watering through the hottest season improves their flowers.

♥ **FLOWER FRIENDS**

SOCIETY GARLIC	DILL
BORAGE	ZINNIA
EUPHORBIA	PENSTEMON
BILLY BUTTON	COSMOS
AMARANTH	AGASTACHE
SHISO	

TIP • Glowing Marigolds are as happy growing in the garden as they are in a planter. A perfect friend to other plants for warding off cheeky pests.

A FROTH OF CLIMBING FLOWERS

✸ WHY
An annual rambling climber with a characteristic flower, bursting with scent and nostalgia. Sweet Peas are deservedly well-loved and a bunch is a seasonal treat in a diversity of colours.

✐ WHAT
Sweet Pea seeds are available as single-coloured varieties or in a happy mix. Though Sweet Peas can be sourced in seedling punnets, the darlings are not fond of having their roots disturbed.

✔ WHEN
Sow in autumn and spring (and even summer in mild climates) for a succession of flowers.

☀ WHERE
Full to partial sun position with good drainage and soil enriched with organic matter. Have a structure prepared for the Sweet Peas to climb, or allow them to ramble in a mound of tendrils and long-stemmed blooms.

❧ HOW
Soak seeds in warm water for 12 to 24 hours (no longer) before planting to soften the hard shell of the Sweet Pea seeds. Sow direct into soil prepared with lots of organic material.

Increase the flower stems by 'pinching out' or 'tip pruning' **(see p. 158)** when plants are about 10 centimetres tall – simply remove the tip to just above the nearest node.

❀❀ VARIETIES
Sweet Pea varieties to grow in an abundance

SWEET PEA 'MATUCANA'
A bright bicolour Sweet Pea of magenta and purple. A full jar of these blooms will saturate any room with a heady and immersive fragrance.

SWEET PEA 'ALMOST BLACK'
From prominent New Zealand Sweet Pea breeder Keith Hammett, this unique specimen is as bold as it is dark – nearly black!

THIS PAGE • Sweet Pea blooms poke out from the climbing vine on straight stems – perfect for a vase.

The fragrance of the old-world and heirloom varieties is so intensely heady. Seek out these varieties if you want the scent permeating your home in spring. 'Matucana', an old-world variety, has the most intense fragrance.

MAXIMUM BLOOMS ON A BUDGET

FULL FLOWER POWER

CHAPTER 3

Bring an abundance of beauty to small spaces and discover the pleasures of pairing flowers with possible friends. There is so much we can do with planters by filling, even layering them with seasonal blooms. These flowers bring the beauty even in limited space.

A PLAYFUL, OLD-WORLD FAVOURITE

✳ WHY

A nostalgic plant. Prolific flowers in a planter or covering the ground.

✎ WHAT

Snapdragons can be grown by seed with a bit of tender, loving care. They can also be sourced as young plants or punnets of seedlings.

⚑ WHEN

Snapdragons are resilient in cooler weather over mild winters, so they can be planted as seedlings and plants in autumn and spring.

☀ WHERE

Full to partial sun. Plant into free-draining soil with added organic matter.

↘ HOW

Young Snapdragons may need protection from slugs and snails but can be surprisingly resilient through extremes in weather.

❀❀ VARIETIES

Together with various colours available in the classic velvety single Snapdragon, 'Madam Butterfly' varieties have ruffled, fuller flowers in a spectrum of pastel and coppery tones. The 'Chantilly' Snapdragon blooms with open faces and mouths agape rather than the usual jaw-like flower form.

♥ FLOWER FRIENDS

CORNFLOWER	AGAPANTHUS
POPPY	PHLOX
ECHINACEA	ACHILLEA
SALVIA	GERANIUM
STRAWFLOWER	AQUILEGIA
STATICE	CERINTHE

TIP • Tip prune back to a leaf joint on all stems on younger seedlings to encourage a bounty of flowers from denser stems **(see p. 158)**.

SPRING FLINGS

✳ WHY Daffodils and Jonquils signal a seasonal change from the depths of winter into the optimism of spring like no other plant.

🖊 WHAT Plant bulbs of Daffodils and Jonquils sourced from quality bulb suppliers. There are many mail-order options while bulbs are dormant.

🍂 WHEN Plant in autumn so the bulb can start growing and become established before the end of winter.

☀ WHERE Full sun to part shade, enrich the soil with organic matter before planting or alternatively use quality potting mix for a planter full of spring delight.

↘ HOW As a general guide, plant the bulb as deep as the bulb is long. As the seasons turn to summer and the flowers are spent, allow the foliage of the Daffodils and Jonquils to yellow as the chlorophyll is drawn back into the bulb. It is tempting to prune off this foliage as it starts to fade but resist the urge until there is no longer any green on the foliage.

♥ FLOWER FRIENDS

HONESTY	*ORLAYA*
POPPY	VIOLETS
FORGET-ME-NOT	BLUEBELLS
HELLEBORE	WINDFLOWERS

TIP • Spring bulbs offer an ideal chance to fill a planter with spring delight for a longflowering display. Once flowering, these planters are easily moved to areas you can see in the garden, or even to be adored inside.

AUTUMN QUEENS

❀ WHY
An old-world plant that is enjoying renewed appreciation for its large and lasting flowers.

✐ WHAT
Plants sourced from quality nurseries. *Chrysanthemums* can also be sourced as seeds.

⚑ WHEN
Plant from autumn to spring so the plant is established before the start of summer.

☀ WHERE
Full-sun position with soil that has been enriched with organic matter or a planter with quality potting mix.

➘ HOW
There is a range of *Chrysanthemum* sizes and the larger ones can be top heavy, so staking or flower supports will help to keep them upright. *Chrysanthemums* can also be grown from cuttings (see p. 154) if you have a garden friend or a particularly lovely one you'd like to see more of.

Trimming or tip pruning your *Chrysanthemum* will encourage more flowers to form.

♥ FLOWER FRIENDS

SALVIA
DAHLIA
CANNA
AGASTACHE
ROSE
SEDUM

EUPHORBIA
VALERIAN
WINDFLOWER
DIANTHUS
CALIFORNIAN POPPY
PELARGONIUM

BRILLIANT ELECTRIC BLOOMERS

✵ WHY
Euphorbias come to cheer with their flowers at the time of the year when we need a floral pick-me-up.

✎ WHAT
Euphorbias are best sourced as plants from nurseries specialising in dry summer plants. Be warned, when they are happy in their conditions, they can prolifically self-seed.

✵ WHEN
As a plant that enjoys dry summer conditions *Euphorbias* do not like to be pruned going into cooler weather. Leave all pruning for spring. When pruning, be careful of their milky white sap, which can irritate skin.

☀ WHERE
Full sun, and tolerant of dry shade positions in the garden. Happy in planters and growing in other vessels.

↘ HOW
Grow in free-draining soil, in the garden or planters. The ground cover *Euphorbias* look especially eye-catching spilling over stone walls.

❀❀ VARIETIES & SPECIES

EUPHORBIA WULFENII
A large, green, mounding *Euphorbia* with electric green flowers.

EUPHORBIA RIGIDA
A ground cover *Euphorbia* with a blush to the spent flowers.

EUPHORBIA ROBBIAE
A shorter *Euphorbia* that can grow well in shade.

EUPHORBIA 'BLACKBIRD'
A *Euphorbia* with moody dark foliage.

♥ FLOWER FRIENDS

KANGAROO PAW	PELARGONIUM	ERIGERON	STATICE
BEARDED IRIS	SALVIA	ERYNGIUM	DIANTHUS
CISTUS	AGAPANTHUS	SEDUM	AGASTACHE

TIP • *Euphorbias* that have self-sown will not necessarily be 'true', meaning the same as their parent plants, so small stem cuttings produce more predictable results if propagating.

RESILIENCE AND BEAUTY IN A BLOOM

✻ WHY
Sedum generally fall into two groups: evergreen ground cover varieties (*Sedum* species); and taller-stemmed herbaceous perennials with large flowers (*Hylotelephium* species).

✐ WHAT
Plants from nurseries that specialise in dry-summer gardens and perennials. *Sedum* can be easily propagated as cuttings **(see p. 154)** during spring.

⚐ WHEN
Autumn through to spring is an ideal time to plant *Sedum* as they are in and emerging from their dormancy. Spring is the perfect time to take cuttings.

☀ WHERE
Full sun and free-draining positions are ideal for *Sedum*.

↘ HOW
Plant into free-draining soil. All kinds of *Sedum* can be grown in planters and grown in composition with other resilient dry-summer species.

✿ VARIETIES & SPECIES

HYLOTELEPHIUM 'MATRONA'
Purple and blue bruised stems with pink dome of flowers.

SEDUM RUPRECHTII
A mounding *Sedum* in dusky pink and grey colours.

HYLOTELEPHIUM 'AUTUMN JOY'
Upright flowers of dusky pink and apricot tones.

♥ FLOWER FRIENDS

AGASTACHE
SCABIOSA
SALVIA

EUPHORBIA
ACHILLEA
ORIGANUM

TIP • *Sedum* flowers are long-lasting when they are at their peak and an enduring cut flower. Ground cover *Sedum* form lovely clustered mounds of flowers and foliage. Taller, herbaceous *Hylotelephium* may even peak in their dormancy, when the flowering stems draw right back to ground level, leaving an incredible stand of delicate, brown stems.

A WONDROUS AND EVOCATIVE FLOWER

✿ WHY With their delightfully dense flowers, *Zinnias* are incredible for their ability to flower brilliantly and prolifically through long periods of dryness without so much as a scowl.

✐ WHAT Grow from seed for a mass of flowers for the garden and to cut and share.

↑ WHEN Sow in late spring to summer. *Zinnia* seeds do need some heat when sown so it is best to sow them into punnets and keep them warm before germination.

☀ WHERE *Zinnias* are sun lovers, enjoying as much sunshine as you can find and a free-draining position.

↘ HOW Regular deadheading will keep *Zinnias* profusely flowering throughout the season.

✿✿ VARIETIES *Zinnias* come in a broad spectrum of colours and forms – double and single blooms of large and small sizes. *Zinnia* 'Queen Red Lime' (pictured opposite) is a personal favourite. The delicious coral, pink and lime tones through this Zinnia are so hard to pass.

♥ FLOWER FRIENDS

SCABIOSA	*CALIFORNIAN POPPY*
VERBENA	*STATICE*
GLADIOLI	*AGASTACHE*
PENSTEMON	*COSMOS*

TIP • A summer annual, *Zinnias* are genuinely the ultimate cut flower – long flowering, long lasting with happy, firm blooms.

BEARDTONGUE

ARCHING STEMS OF JOY

❋ WHY	*Penstemon* have dreamy, long arching steps of bold tubular flowers that hang down and open like elongated trumpets. The common name, Unusual Beardtongue, is evocative of the occasionally fuzzy stamens some varieties have inside.
🌰 WHAT	Plants can be sourced from nurseries that specialise in perennials, and it is possible to grow them from seed from a specialty supplier.
🌱 WHEN	Sow seeds into punnets to grow from spring to early summer. Plants can be sourced and planted in most seasons, apart from the full heat of summer.
☀ WHERE	Full sun and free-draining position. Improve the soil with organic matter before planting.
✋ HOW	*Penstemon* enjoy soil which has had organic matter added to it, or quality potting mix if in a planter. *Penstemon* are prolific with their flowering stems and flower for the better with some extra water over summer.

VARIETIES & SPECIES

PENSTEMON STRICTUS
Resilient for dry periods through summer with a bright purple flower.

PENSTEMON BARBATUS 'COCCINEUS'
Has lovely stems and coral-red flowers and, though finer in form, is far more resilient than similarly coloured cultivars.

PENSTEMON 'SOUR GRAPE'
Bold tones of blue in this grape-coloured flower.

PENSTEMON 'RAVEN'
A rich plum *Penstemon*.

PENSTEMON 'FIRE BIRD'
Less fire and more a rich coral with even darker stems. Completely delicious.

FLOWER FRIENDS

KANGAROO PAW
AGAPANTHUS
BEARDED IRIS
ECHINACEA
SALVIA

NICOTIANA
DAHLIA
SCABIOSA
COSMOS
AGASTACHE

THIS PAGE • *Penstemon* 'Raven' with flower friend *Salvia* 'Amethyst' (see p. 94).

TIP • If you cut back *Penstemon* after their first flush of flowers they will often produce a second flush. These are great flowers to enjoy both in small spaces and as cut flowers inside. *Penstemon 'Sour Grapes'* pictured.

CONEFLOWER

A TRUE DARLING OF SUMMER

✸ WHY

A hero of a flower with proud centres that point upwards and petals that flex downwards, creating the visible skirt that gives them the distinctive 'coneflower' common name.

✐ WHAT

Seeds are available to grow as single species or a mix of colours. Plants of the *Echinacea* are available between spring and autumn when herbaceous perennials are most available.

✦ WHEN

Sow seeds, plant and divide in early spring. Seeds enjoy warmth to get started so consider a warm position to encourage them to germinate.

☀ WHERE

A full-sun position with well-prepared, free-draining soil. Cluster several together, planting closely if using a planter. The extra density will give you a big show of brilliant cone-like flowers.

➘ HOW

Seeds should be planted into trays, and germination is faster and more successful with a bit of base heat. It is easy to save the seeds of coneflowers. Ensure the seed heads are dry and mature before cutting from the stems.

❀ SPECIES

ECHINACEA PURPUREA
Displays glorious hues of sun-saturated colours. Varieties and cultivars of Echinacea purpurea have firm, solid, structural flowers that pose proudly. The petals flex out and slightly downwards.

ECHINACEA PALLIDA, ECHINACEA PARADOXA, ECHINACEA SIMULATA
While Echinacea purpurea have dense, proud, upright flowers,

I prefer the delicate whimsy of other species. Their form is taller with the same habit, often with tap roots drilling down into the soil. These Echinacea still have the coneflower centre they are revered for, but rather than a sun-shaped form with rays of petals flexing out, they have gentle hanging petals like a skirt made of fabric strips. The petals catch the slightest bit of wind and give the appearance of a jellyfish.

♥ FLOWER FRIENDS

SCABIOSA	*DAHLIA*	*RUDBECKIA*	*AGASTACHE*
SEDUM	*COSMOS*	*SALVIA*	*PHLOMIS*

TIP • It is fabulous to observe *Echinacea* flowering in the garden. A small, round disc rises on a thick stem from the strappy basal foliage, gaining height quickly once it reaches a particular elevation. Petals start to peek out from the disc, and over the coming days, they flow out and down away from the cone.

THE ULTIMATE BEAUTY QUEENS

✻ WHY　The abundant flowers of the *Dahlia* are so prolific and generous from summer into autumn. It is a special delight that their magnificence has been discovered by a whole new generation of beauty seekers.

✐ WHAT　Tubers can be sourced in winter to plant in spring. Many *Dahlia* growers will take pre-orders as early as late summer. Ensure you source from reputable suppliers to avoid the *Dahlia* virus. If concerned, source *Dahlias* as plants. That way, you can tell if the foliage is showing signs of the virus (marked, deformed or discoloured) and avoid purchasing. Plants are available from spring to early autumn.

⚑ WHEN　Plant tubers in spring. Some gardeners wait until the eyes or new shoots of the *Dahlias* are obvious before planting.

☀ WHERE　Full sun into free-draining soil improved with organic matter.

☟ HOW　Plant tubers on their side with their neck or growing eyes just above the ground level and then water them in well once the tips start to grow. The taller varieties with massive flowers can become top-heavy due to the lush stems and prolific flowers. It is good to support these plants with staking.

❀❀ VARIETIES　Where to begin? The list is endless, and collecting the different flower shapes, colours and sizes can become quite the addiction.

♥ FLOWER FRIENDS

SCABIOSA　　　　*SALVIA*
UMBELLIFERS　　*ECHINOPS*
COSMOS　　　　*NICOTIANA*
SUNFLOWER　　　*CANNA*
RUDBECKIA　　　*CROCOSMIA*

THIS PAGE • *Dahlias* make such amazing cut flowers. The blooms are charming and provide long-lasting wonder inside and in the garden. They range from enormous, plate-sized with fluted or informal petals to small, button-sized pom-poms.

AN UPTIGHT DELIGHT

WHY Long, slender stems rise from lush rosettes of foliage dripping with long tubular flowers.

WHAT Foxgloves can be grown from seed and also sourced as young plants to fill planters and tight pockets of the garden where some flowering height would be just right.

WHEN If you plant biennial Foxgloves in autumn, they will flower in spring within the first year. The first set of flowers will be the largest but smaller flower spikes can emerge in summer after deadheading. Foxgloves can be grown quickly from seed. Germinate in the autumn in punnets and transplant into the garden.

WHERE Plant Foxgloves in full sun to partial shade in fertile, free-draining soil.

HOW The lush foliage can be tempting for slugs, snails and caterpillars. Keep an eye on them and consider natural treatments if the damage is past your tolerance.

FLOWER FRIENDS

- **ROSE**
- ***SALVIA***
- ***PELARGONIUM***
- ***SCABIOSA***
- **SWEET PEA**
- **SNAPDRAGON**
- **POPPY**
- **BEARDED IRIS**
- ***CALENDULA***
- ***NIGELLA***
- **HOLLYHOCK**
- ***ECHINOPS***
- **UMBELLIFERS**
- **SHASTA DAISY**

TIP • Foxgloves make an impact when several are in bloom simultaneously, either in a big sweep or in smaller clusters dotted through the garden. There are many species to consider, from earthy shades of cream to coral, and flower spikes ranging from fine to bold.

ENDURING BEAUTY

CHAPTER 4

The extremes of the climates we live in should not exclude us from the possibility of plants. It's a rare garden or patch or balcony somewhere that has the perfect climatic conditions and natural rainfall. These beauties are powerful in their resilience in the face of a changing climate.

FEVERFEW, MARGUERITE AND ERIGERON DAISY

CAREFREE, SIMPLE BEAUTY

❋ **WHY**　These blooms are great cut flowers and bring so much cheer when shared.

✿ **WHAT**　Plant young and advanced plants. Source these flowers from a grower that specialises in dry-tolerant plants.

✦ **WHEN**　Most daisies can be planted year-round but it's best to avoid any planting in the peak of summer if you live in a dry-summer climate.

☀ **WHERE**　Full sun in free-draining soil.

✎ **HOW**　Herbaceous perennial daisies benefit from cutting back after they retreat for their dormancy. All evergreen and shrub daisies will benefit from a trim after flowering to keep the foliage dense and full and to encourage further blooming. *Erigeron* are resilient during dry summer periods but may need extra watering to encourage flowering when establishing.

❀ **VARIETIES**

RHODANTHEMUM
The genus of Moroccan daisy has sweet daisy blooms over mounds of finer foliage.

MARGUERITE
This nostalgic and simple daisy is available as an evergreen perennial or a shrub. It is the perfect bloom for making daisy chains.

ERIGERON
Seaside daisies or Beach Asters have fine, thin petals in shades of white or pink on delicate foliage. *Erigeron karvinskianus* 'LA Form' has large flowers in a dreamy pastel pink. *Erigeron glaucus*, known commonly as Beach Aster, is an excellent ground cover and weed suppressant with pale pink and lilac flowers.

♥ **FLOWER FRIENDS**

SALVIA
DIANTHUS
CALIFORNIAN POPPY
SEDUM
ERYNGIUM
STATICE
PELARGONIUM

KANGAROO PAW
POPPY
SCABIOSA
AGASTACHE
PENSTEMON
AGAPANTHUS

CHAPTER 04

MAGICAL WHIMSY

✳ WHY Umbellifer is a general description of any flower with a wonderful, distinctive umbrella shape. Small flowers form tiny branchlets with clustered florets that rise upwards together, creating the impression of a dome. The little florets form a mini version of the overall bloom. Super cute!

✐ WHAT Umbellifers tend to have tap roots and therefore don't enjoy having their roots disturbed, so growing from seed is best. Sow directly into the ground.

⚐ WHEN This loose group of umbellifers are all annuals resilient for the cooler months and therefore can be sown from autumn.

☀ WHERE Full to partial sun in free-draining soil but tolerant of different soil types.

⬂ HOW False Queen Anne's Lace and Flowering Carrots can become large clumps of stems and flowers arching and spilling over their neighbours; it is best to stake or bind them and thin them out as they start picking up size and speed.

♥ FLOWER FRIENDS

DAHLIA	VERBENA
SCABIOSA	RUDBECKIA
COSMOS	SALVIA
FOXGLOVE	KANGAROO PAW

TIP • The Umbel group (all in the family of Apiaceae) is expansive and includes carrot, fennel, dill and False Queen Anne's Lace. Flowering Carrots can be found in the most whimsical shades of chocolate or pink. Umbellifers are usually grown as annual plants but many will reshoot and become perennial with a regular cutback.

GLOBE FLOWER

A LUMINESCENT GLOBE

✳ **WHY**
Perfectly round and spiky orbs, with flowers ranging in colour from silver to bright blue, take on a luminescence in the early morning and late evening light. The seed heads fade to silver and grey as they age.

✿ **WHAT**
Because seed germination can be unreliable, *Echinops* are best sourced as plants. You will find them available as herbaceous perennials from spring to autumn.

✦ **WHEN**
It is ideal to plant these perennials in autumn, allowing for time over winter for them to become established. This is a perennial that can self seed in the garden.

☀ **WHERE**
Full sun, partial shade. They will thrive in low-nutrient soil but must have good drainage.

🗤 **HOW**
Echinops look amazing when planted in clusters. The plants can become top-heavy, so they benefit from staking. Deadhead by cutting back to the next node or leaf joint to encourage continued flowering. Cut plants back to the base in late autumn, and new foliage will appear in spring.

❤ **FLOWER FRIENDS**

RUDBECKIA	*ERYNGIUM*
AGASTACHE	*AGAPANTHUS*
ECHINACEA	*COSMOS*
GEUM	*SCABIOSA*
BERGAMOT	*PELARGONIUM*

TIP • Once *Echinops* are established, you can divide the clump and transplant them throughout the garden or pot them for friends. Do this at the start of spring when the plants emerge from dormancy.

SALVIA NEMOROSA, SALVIA VISCOSA, SALVIA APIANA, SALVIA DISCOLOR,
SALVIA 'RUBIN', SALVIA TRANSSYLVANICA, SALVIA CORRUGATA

A FLASH OF ADORATION

✻ WHY   If only allowed one group of plants, you must choose *Salvia*.

✐ WHAT   Seeds, seedlings and plants. Varieties that are harder to find may first need to be grown from seed. There is a glorious array of *Salvias* to grow and all with varied tolerances; many are resilient to dry summers, some more than others.

↟ WHEN   Plant *Salvias* from autumn through to spring. *Salvias* vary in their form and life cycle being herbaceous perennials, evergreen perennials or annuals here for a few seasons. If growing annual *Salvias* from seed, spring planting is ideal.

☀ WHERE   Full sun, some species tolerate shade but will have reduced flowering. Herbaceous perennial varieties are usually frost tolerant. Plant into free-draining soil.

↘ HOW   Tip prune at the start of spring to increase flowers and deadhead spent flowers through the season. Herbaceous perennials can be cut back to their base after flowering, which sometimes encourages another flush of flowers before the season ends. The woodier *Salvias* can be cut back by a third, maybe even more, to the leaf joints to encourage vibrant foliage and flowers the following season.

♥ FLOWER FRIENDS

AGASTACHE	*CALIFORNIAN*	*ERYNGIUM*	*DIANTHUS*
COSMOS	*POPPY*	*SEDUM*	*PENSTEMON*
EVERLASTING	*BLUE LACE*	*EUPHORBIA*	
KANGAROO PAW	*FLOWER*	*SCABIOSA*	
PELARGONIUM	(see Umbellifers)	*BEARDED IRIS*	

TIP • *Salvias* are great plants to practise propagating from cuttings **(see p. 154)**. Herbaceous perennials can easily be divided once they have established. It is best to do this in early spring, when the perennials have started to sprout, or in autumn so the plants can establish through the winter months and are primed for a long flowering season ahead.

THIS PAGE • *Salvia* flower stems stretching to the sun.

SEA HOLLY

A SCULPTURAL DELIGHT

✳ WHY　Fascinating spiny flowers shoot up from a rosette of foliage at the ground level. The flowers carry a metallic sheen that is almost other-worldly.

◔ WHAT　Seeds and plants can both grow in the garden. Source different species and cultivars from nurseries specialising in perennials and plants suitable for dry summer climates.

⚐ WHEN　Sow seeds in autumn or spring, plant at any time but extra water will be required if planting in summer. Best to avoid if you can.

☀ WHERE　Plant in a free-draining, even sandy, soil. Some *Eryngiums* are tolerant of frost but are not happy in wet and waterlogged soil, especially in cooler locations.

❧ HOW　*Eryngium* species can be tricky to germinate. So start with a plant and try seeds once the plants are flourishing in your garden.

❀❀ VARIETIES & SPECIES

ERYNGIUM 'SILVER SALENTINO'
Silver flowers that are illuminated in low light.

ERYNGIUM OVINUM
Australian *Eryngium* growing a mass of airy flowers with a spiky, pointy form.

ERYNGIUM 'MISS WILLMOTT'S GHOST'
Silvery-green flowers with a bract forming a prickly collar.

ERYNGIUM PLANUM
Sea Holly flowers of metallic blue.

♥ FLOWER FRIENDS

SALVIA
CALIFORNIAN POPPY
GLADIOLI
NICOTIANA
PELARGONIUM
SEDUM

WINGED EVERLASTING
UMBELLIFERS
COSMOS
SCABIOSA
KANGAROO PAW
BILLY BUTTON

ANISE HYSSOP, HUMMINGBIRD MINT

A GENEROUS SUMMER FAVOURITE

✿ **WHY**
Fragrance fills the air when you brush past this aromatic perennial.

✐ **WHAT**
Seeds and plants from nurseries that specialise in perennial plants.

⚘ **WHEN**
Sow seeds direct in early spring; at the same time it is also possible to propagate cuttings (see p. 154). For plants, plant in spring and autumn outside of the hottest and coldest months.

☀ **WHERE**
Full sun with free-draining soil.

🜸 **HOW**
Deadhead tired flower stems and cut back, leaving a section of stem and foliage in preparation for the plant's dormancy during winter. *Agastache* can be propagated from cuttings or by dividing larger clumps.

❀ **VARIETIES**

AGASTACHE 'BLUE BOA'
Double flower stems of a rich and appealing blue.

AGASTACHE 'SWEET LILI'
A resilient variety with spikes of delicate purple flowers.

AGASTACHE 'BLUE FORTUNE'
Dense purple-blue flowers.

AGASTACHE 'LEVI'
Airy stems of butter-yellow flowers.

♥ **FLOWER FRIENDS**

EUPHORBIA
ERYNGIUM
PENSTEMON
SALVIA
ROSE

SCABIOSA
BILLY BUTTON
STRAWFLOWER
SEDUM
PELARGONIUM

TIP • There are two main groups of *Agastache*. Those from Asia and North America are very structured and have dense, double-flowering stems held upright, creating robust vertical markers in the garden. This group needs regular watering during dry periods. The second, more resilient group is twiggier and airier. They produce long, floriferous stems of delicate tubular flowers that glow in the sunlight.

THIS PAGE • The fine, tubular flowers of Agastache 'Apricot Sprite'.

ABUNDANT SUMMER BLOOMS

✻ WHY *Cosmos* bring their easy, airy blooms to summer, surrounded by fine, feathery foliage. Their happy daisy-like flowers are open and swing gently in any garden breeze.

✐ WHAT Easily grown from seed in a range of wonderful colours. *Cosmos* seedlings are also available seasonally from nurseries. Plants of the perennial Chocolate *Cosmos* (*Cosmos atrosanguineus*) can be sourced through nurseries and online from perennial specialists.

✇ WHEN Sow seeds in early spring.

☀ WHERE Full sun in a free-draining position.

❧ HOW Give *Cosmos* plenty of care and preparation to get them started, though they can surprise by thriving in low-nutrient soils. As *Cosmos* grow, they will be happily independent, doing their thing into the warmer months. Cutting the flowers to bring into your home, or deadheading spent flowers, will encourage more and more successive buds throughout the season.

♥ FLOWER FRIENDS

DAHLIA	*AGASTACHE*
SCABIOSA	*SALVIA*
SUNFLOWER	*ECHINACEA*
NICOTIANA	*RUDBECKIA*
DIANTHUS	*SEDUM*
BILLY BUTTON	*STATICE*
STRAWFLOWER	*VALERIAN*
KANGAROO PAW	*EUPHORBIA*
ERYNGIUM	*AGAPANTHUS*

TIP • Follow your heart and experiment with the flower and colour combinations that appeal to you. And remember, as much as we love *Cosmos*, bees, butterflies and beneficial insects adore them even more.

STRAWFLOWERS AND PAPER DAISIES

AN EASY SEA OF BEAUTY

❋ WHY Petals dry as paper that make a distinct rustle as you run your fingers around the flowers. Everlasting is the common name for a varied group of different plants with the same charming qualities.

◔ WHAT Many varieties can be sourced as seed. Nurseries specialising in plants suitable for dry summer climates often stock different cultivars, especially from the *Xerochrysum* and *Chrysocephalum* genera.

⚑ WHEN Sow seeds in autumn through to spring. Plants can be planted from autumn through to spring.

☀ WHERE All Everlastings love a full-sun position.

❧ HOW Everlastings are not fussy when it comes to improved soil or quality potting mix. They don't necessarily need an Australian potting mix, but free-draining soil with organic matter is a must.

✿❀ VARIETIES

STRAWFLOWERS
Formerly *Helichrysum*, now *Xerochrysum bracteatum*, an abundance of large flowers in brilliant mixes or lovely, single jewel and pastel colours. A beautiful, enduring cut flower, easy to dry.

**PAPER DAISY
(*Rhodanthe chlorocephala*)**
Comes in white and pink shades and is perfect for meadows or planting in large drifts.

♥ FLOWER FRIENDS

CALIFORNIAN POPPY
STATICE
COSMOS
SCABIOSA
CALENDULA

CORNFLOWER
POPPY
NICOTIANA
PELARGONIUM
SEDUM

THIS PAGE • The ribbed stems of the Winged Everlasting (*Ammobium alatum*).

CHAPTER 04

TIP • With firm and iridescent petals like shiny, jewel-coloured beetle shells, these flowers are partially pre-dried and everlasting. Harvest them when they are in bud or have just opened to dry flowers and preserve their beauty. Cut long stems – the longer, the better. Remove any excess foliage. Tie small bunches lightly and hang them upside down, out of the sunlight in a well-ventilated place.

SPRIGHTLY FLOWER SPIKES

✻ **WHY**	Reclaim this old beauty. As a cut flower Gladioli are quite striking and the species is full of treasures to enjoy.	

✿ **WHAT** Bulbs can be sourced in pots or as bare root bulbs.

🌱 **WHEN** Bulbs planted in autumn to spring for summer flowering.

☀ **WHERE** Full-sun positions in the garden with shelter from damaging winds.

✎ **HOW** *Gladiolus* species need free-draining soil and will flower with more profusion with supplementary watering during hot periods. Species *Gladiolus* are resilient for drier conditions.

♥ **FLOWER FRIENDS**

RUDBECKIA	*KANGAROO PAW*
STRAWFLOWERS	*WAHLENBERGIA*
UMBELLIFERS	*NASTURTIUM*
DAISY	*CANNA*
FORGET-ME-NOT	*EUPHORBIA*

TIP • *Gladiolus* species often referred to as dwarf or Nana Gladioli are smaller and shorter in form and flower compared to the large cut flowers popularised by Dame Edna. Both can be grown in the garden with options for tall or short Gladioli.

BLACK-EYED SUSAN

BOLD SUMMER BLOOMS

✿ **WHY**	A resilient and prolific summer bloomer, *Rudbeckia* are bright and bold but especially abundant for the summer garden.	

WHAT	Seeds can be sourced for *Rudbeckia* from specialty suppliers. Plants are available from quality nurseries from spring to summer.	

⚐ **WHEN**	Sow seeds in spring with the benefit of the warming season. *Rudbeckia* can vary in their resilience for the cold so it is best to plant in the spring.	

☀ **WHERE**	Full-sun position with free-draining soil.	

✋ **HOW**	Early tip-pruning will increase the abundance of flowers, and deadheading will continue the flowering into early autumn. As clumps of the *Rudbeckia* expand each year, they can be lifted and divided in spring.	

VARIETIES

RUDBECKIA HIRTA
Better grown as an annual in some locations. Varieties include 'Cherry Brandy', with rich burgundy and chocolate flowers, and 'Marmalade', with tightly held, rich-yellow petals with small prominent cones – so abundant, brilliant and bold – a fantastic cut flower.

RUDBECKIA MAXIMA
Glorious, bold grey-green leaves and yellow petals surrounding a proud, cone-shaped centre.

RUDBECKIA LACINIATA
A tall and slender beauty with very long flower stems and smaller leaves. The light-yellow flower discs dance on the top of the foliage.

♥ **FLOWER FRIENDS**

SALVIA
AGASTACHE
SCABIOSA
DAHLIA

AMARANTH
ARTICHOKE
CANNA
SUNFLOWER

TIP • All *Rudbeckia* can be used as cut flowers but the cultivars of *Rudbeckia hirta* are the most abundant and the best. The varieties flower profusely on long, strong stems and give the appearance of a flowering dome.

THIS PAGE • The cherry-chocolate tones of *Rudbeckia* 'Cherry Brandy'.

SELF-SUFFICIENT BLOOMS

CHAPTER 5

In the busy modern world we can still have beauty around us by choosing flowers that are both abundant and happy to tend to themselves.

DIANTHUS

CARNATION
UNREALISED BEAUTY

❋ WHY Unfairly derided as the common carnation, the *Dianthus* is set for a comeback. With a resilient bloom and a long flowering season, it feels perfect to use throughout gardens to bring beauty.

✐ WHAT Seeds and plants can be sourced from seed suppliers and quality nurseries. It is possible to also propagate *Dianthus* from cuttings **(see p. 154).**

✦ WHEN Sow seeds direct in spring to early autumn and plant *Dianthus* year-round, avoiding the peak of heat in summer.

☀ WHERE Full sun with excellent drainage.

◥ HOW Deadheading will encourage the plants to rebloom their happy flowers.

♥ FLOWER FRIENDS

ERIGERON	*EUPHORBIA*
ERYNGIUM	*VALERIAN*
SEDUM	*SALVIA*
GLADIOLI	*AGASTACHE*
BEARDED IRIS	*SCABIOSA*

TIP • Given their tolerance for dry positions, there seems to be a big opportunity for combining *Dianthus* with other dry-tolerant plants outside the usual 'cottage garden' style. Experiment away!

STATICE

LIMONIUM SINUATUM, LIMONIUM PEREZII, LIMONIUM PEREGRINUM, PSYLLIOSTACHYS SUWOROWII

SEA LAVENDER

A SEA OF PAPER PETALS

WHY A common name for a robust group of plants sharing the paper-like texture in their flowers and resilience for dry summer conditions. Statice flower on effortlessly.

WHAT Source as seeds or plants from nurseries and seed suppliers.

WHEN Statice can be planted year-round with the seeds ideally sown in spring.

WHERE Full sun and free-draining position.

HOW Propagate perennial plants by division in spring. Cut the stems at the base just as they begin to flower and hang them to dry.

VARIETIES

COMMON STATICE
(*Limonium sinuatum*)
Several tall, long stems ending with a small cluster of flowers, commonly used as cut flowers and grown as an annual plant.

RUSSIAN STATICE
(*Psylliostachys suworowii*)
A riot of tall, fluffy, pink flowers on long, fine stems.

PINK STATICE
(*Limonium peregrinum*)
Clusters of delicate, pink flowers atop stems with oval, leathery leaves.

SEA LAVENDER
(*Limonium perezii*)
Basal clump of earthy leaves; big clouds of flowers on thin, branching stems.

FLOWER FRIENDS

ECHINACEA	*SALVIA*	*COSMOS*	EVERLASTING
SEDUM	*PELARGONIUM*	*SCABIOSA*	*EUPHORBIA*
AGASTACHE	BEARDED IRIS	BILLY BUTTON	KANGAROO PAW

TIP • The papery nature of the flower lends itself perfectly to preserving their beauty by drying the bloom. Hang upside down in a dry, well-ventilated location for a period before arranging.

QUEEN OF THE SPRING GARDEN

✽ WHY

Bearded Iris are so glorious and underrated. Their flowering time is short and swift, but they make such a statement that they deserve a place in every garden.

✐ WHAT

Bare-root rhizomes. Purchase in autumn, selecting rhizomes that look nice and chunky, not sad and withered. Divide rhizomes after flowering and when growth has slowed in autumn.

↑ WHEN

Late summer to early autumn. Plant after you receive the rhizome, a quick soak in a weak seaweed solution before planting will encourage them on a good growing start.

☀ WHERE

Full sun with free-draining soil. They may be content with only afternoon sun if you want to try.

➘ HOW

There are different perspectives on how to plant the rhizome. A section of the rhizome can be exposed to the sun by sitting on the top of the soil or planted shallowly under a fine layer of soil. Both techniques work in temperate climates, but leaving the rhizome exposed may be more important in cool temperatures as they enjoy a bit of heat. It also allows for extra drainage in places that are wet.

♥ FLOWER FRIENDS

POPPY
CALENDULA
SALVIA
EUPHORBIA

STATICE
BLUE LACE FLOWER
CALIFORNIAN POPPY
VALERIAN

THIS PAGE • Ink-stained details of Bearded Iris petals as they begin to fade.

SELF-SUFFICIENT BLOOMS

LILY OF THE NILE

FLORAL FIREWORKS

❋ **WHY**
Though we can be cautious of the 'Aggies' who can act like thugs, there are many modern Aggies that won't go rampant and still provide a display of flowers with ease.

✿ **WHAT**
Plants can be sourced from quality nurseries and sometimes as bare root rhizomes.

⚘ **WHEN**
Agapanthus can be planted from autumn to spring.

☀ **WHERE**
Full and partial sun in free-draining soil.

✎ **HOW**
Ensure you have sterile varieties or remove the flower heads before seeding and avoid planting near bushland. Additional watering through summer will improve flowering, but *Agapanthus* are very dry-hardy plants. They are also excellent in planters. As any herbaceous varieties retreat into dormancy over the cooler seasons, cut off the spent foliage at the base of the plant. Deadhead the spent flowers regularly. Divide after flowering or in early spring.

♥ **FLOWER FRIENDS**

PELARGONIUM	ERYNGIUM
KANGAROO PAW	SEDUM
BILLY BUTTON	SALVIA
SCABIOSA	STRAWFLOWER
COSMOS	BEARDED IRIS
RUDBECKIA	AGASTACHE

TIP • Old evergreen varieties of *Agapanthus* deserve their reputation as garden bullies and environmental pests. However, modern, sterile varieties and herbaceous perennials whose growth is controlled by winter dormancy offer a guilt-free and responsible way to bring long-stemmed white, blue and inky-purple explosions to your summer garden.

CAREFREE SENSIBILITIES

✱ WHY — These ground covering poppies have happy, open flowers and petals with a gossamer silk texture. Multi-dimensional, their foliage intermeshes and knits between other plants or forms low mounds when not interceded by plant neighbours.

✎ WHAT — Packets of single or mixed colour seeds can be sourced.

⚑ WHEN — Sow seeds in late summer to spring, even year round in frost-free areas.

☀ WHERE — Full sun. Grows well in containers, is wonderful as an annual underplanting and a great way to fill gaps.

🢆 HOW — Sow seed directly in free-draining soil. In temperate climates, Californian Poppies will bloom through the cooler months and into spring and summer. They will self-seed, so you can edit, transplant or just let them do their thing.

❀❀ VARIETIES

'ROSE CHIFFON'
Double Californian Poppy with a delicious bifold detail to the petal.

'PURPLE GLEAM'
A soft lilac open Poppy that flowers summer long.

'SUNDEW'
A sweet and small butter cream flower that disperses through the garden.

♥ FLOWER FRIENDS

BEARDED IRIS	PELARGONIUM	STATICE	SALVIA
ERYNGIUM	EUPHORBIA	AGAPANTHUS	AGASTACHE
KANGAROO PAW	ALOE	BILLY BUTTON	SCABIOSA
SEDUM	DIANTHUS	RUDBECKIA	

TIP • When the conditions are right for a super bloom event, the hills of California become coated with flowering poppies. They flower in such abundance that the landscape changes colour.

A PROLIFIC PERENNIAL

✿ **WHY**
Valerian will take care of itself through dry summers, bringing joy in the sculptural blooms. From a distance, the clump of flowering stems can look like frothy clouds of colour.

✐ **WHAT**
Valerian can be grown from seed. Source plants for your garden from nurseries that specialise in stock for dry-summer climates.

⚐ **WHEN**
Germinate the seeds in spring. Valerian can be planted year-round avoiding the hottest time of the year.

☀ **WHERE**
Full sun and free-draining position.

➤ **HOW**
Valerian is easy to grow; it thrives in alkaline soil improved with organic matter, but tolerates different soil compositions. It is not very long-lived but will readily self-seed in warmer climates. You can cut the flowers back before they seed or leave a few volunteers to emerge as seedlings.

❤ **FLOWER FRIENDS**

SALVIA
PELARGONIUM
EVERLASTING
COSMOS
BEARDED IRIS
SEDUM

KANGAROO PAW
ERYNGIUM
SCABIOSA
BILLY BUTTON
BLUE LACE FLOWER

PINCUSHIONS

FLUFFY, FLOATING FLOWERS

| ✴ **WHY** | *Scabiosa* form cute circular buds that resemble floating pincushions on stems before opening into gorgeous blooms. |

| ✐ **WHAT** | Seeds and plants. |

| ⚑ **WHEN** | Sow seeds in spring either directly or into punnets and transplant seedlings. |

| ☀ **WHERE** | Full sun with free-draining soil. |

| ➤ **HOW** | Transplant seedlings carefully without disturbing the roots. Top-heavy pincushion *Scabiosa* will benefit from staking. Plants go into dormancy over the cooler months. Cut them back to roughly 10 centimetres above the soil and they will get started once again as the earth warms in spring. |

✿ VARIETIES

SCABIOSA ATROPURPUREA
Produces an abundance of tall, fine stems topped with floating flowers that resemble tiny UFOs. Great intermingled with other tall perennials and annuals.

SCABIOSA CAUCASICA
Large striking flowers with a face rimmed by delicate petals in shades of white and pastel-blue to mauve. Long, sturdy stems make this an excellent cut flower.

♥ FLOWER FRIENDS

STRAWFLOWER
KANGAROO PAW
RUDBECKIA
COSMOS
SUNFLOWER

DAHLIA
AGASTACHE
SNAPDRAGON
SALVIA
PENSTEMON

TIP • *Scabiosa* are prolific self-seeders and are considered weeds in some parts. Care and consideration are necessary when growing them in your area, especially if you live near bushland.

SELF-SUFFICIENT BLOOMS

139

ORNAMENTAL TOBACCO

A DREAMY CHARMER

✸ WHY   Ornamental Tobacco or Flowering Tobacco is grown for its lovely, dreamy flowers rather than for leaves to dry as smoking tobacco.

✿ WHAT   Tobacco plants are restricted in some jurisdictions. Seeds and small plants can be sourced from specialist growers if legally available.

⚐ WHEN   Sow seeds direct or plant seedlings from autumn to spring.

☀ WHERE   Full sun to partial shade into soil enriched with organic material.

✍ HOW   *Nicotiana* is grown as an annual but can act like a perennial in temperate climates. Cut flowers back to the rosette base after flowering. Collect the tiny granular seeds from the sticky, cup-shaped seed capsule.

✿✿ VARIETIES

NICOTIANA LANGSDORFFII
Striking lime-green, tubular flowers.

NICOTIANA ALATA
Open faces of pale green stars.

NICOTIANA SYLVESTRIS
A large cultivar available in various colours. The long stems of the white-flowered variety are useful for lighting up shaded nooks.

♥ FLOWER FRIENDS

ECHINACEA
DAHLIA
COSMOS
SCABIOSA
SALVIA
AGASTACHE

PELARGONIUM
ERYNGIUM
FOXGLOVE
POPPY
BEARDED IRIS

TIP • Dreamy, tubular flowers hang downwards, along arching arms, their faces dangling from the curved stem. This charming, old-world plant, familiar from our grandmas' gardens, attracts butterflies that love to feed on their nectar.

THE REIGNING MONARCHS

✱ WHY
Roses are the reigning monarchs of ornamental gardens and a lovely flower to grow.

✏ WHAT
Source plants in pots or as bare root roses (loose roses without potting mix and a pot while the rose is sleeping during its winter dormancy). Pre-orders are usually taken before the bare root season.

🌱 WHEN
Buying bare root plants is one of the most economical ways to source roses for your garden. It is also the easiest way to get your hands on a wide range of roses through mail order.

☀ WHERE
Full sun (some ramblers and climbers are tolerant of partial shade) in fertile, free-draining soil.

🍃 HOW
Roses are deciduous and, depending on the climate, most will lose all their foliage in the cooler months; this is the time to prune the plant into an open form to reduce the risk of fungal diseases and promote flower production. When pruning and cutting flowers, cut the stem at an angle just above the leaf joint. Cut flowers just before the bloom fully opens for a long-lasting cut flower.

❀ VARIETIES

ROSA 'MUNSTEAD WOOD'
Crimson blooms and a rich fragrance.

ROSA 'SUMMER SONG'
Unusual coral-coloured flowers.

ROSA 'MUTABILIS'
Rambling and climbing varieties with single, open flowers in a spectrum of peach, copper and pink tones.

♥ FLOWER FRIENDS

SALVIA
CALIFORNIAN POPPY
NASTURTIUM
DIANTHUS
POPPY
GLADIOLI
SCABIOSA
CATMINT

AGAPANTHUS
SEDUM
PENSTEMON
AGASTACHE
FOXGLOVE
NICOTIANA
BEARDED IRIS

AN AUSTRALIAN ICON

✻ WHY
The self-sufficient clumps host the most incredible flower stems filled with the velvet 'paws' of the kangaroo. When flowering, the long, thin stems create a see-through quality, blending the foreground and background together.

✎ WHAT
Source plants as tubestock or advanced plants from specialist nurseries stocking Australian and dry-summer plants.

🌱 WHEN
Sow seed in spring, summer and autumn; plant potted and tubestock at any time.

☀ WHERE
Full sun in gritty or sandy free-draining soil.

🖐 HOW
Use a fertiliser designed for Australian plants (low in phosphorus) and ensure plants get enough moisture while establishing and during the summer months. After a year or two, and once Kangaroo Paw is well established, they are resilient and respond well to a hard prune of their foliage after flowering. As the clumps of the Kangaroo Paw increase, they can be divided in the cooler months.

♥ FLOWER FRIENDS

BLUE LACE FLOWER
AGASTACHE
COSMOS
VALERIAN
RUDBECKIA
SEDUM
SALVIA

EUPHORBIA
PENSTEMON
DAISIES
STATICE
WINGED EVERLASTING
BILLY BUTTON
SCABIOSA

METHODS FOR MAXIMUM FLOWERS

CHAPTER 6

This chapter outlines basic techniques to maximise flowers in your garden by saving and growing seeds, and ways to make more plants by dividing and growing from plant cuttings.

HOW TO SAVE SEEDS

For a seed to be viable you need to ensure that it has matured or ripened and is stored correctly. Collect seeds once they have matured on the plant, after the flower has waned. Wait until the seed is completely dry before collecting.

Always collect seeds from the happiest and healthiest plants in your garden. Avoid collecting seeds from plants that look stressed, unhappy, diseased or pest ridden.

CLEANING SEEDS

Once the flower heads or pods have been collected, allow them to dry for a fortnight in a well-ventilated and dry location. Next, separate them from the remaining plant material such as dried petals and husks. This can be done by hand, or sieves can be useful to remove any debris.

STORING SEEDS

Store seeds in clean paper – a new envelope or small paper bag, even using a folded piece of recycled paper to make a pocket.

Label it with the name, your favourite notes and the date.

It is important to keep the seeds at a stable and consistent temperature with low humidity. This may just be in a cupboard or room of your house which fluctuates the least throughout the day and the year, away from any appliances or heating or air conditioning vents.

UV will degrade seeds over time and light will encourage germination as well as influence the temperature. It is best to store your seeds in a dark spot out of direct sunlight.

SOWING SEEDS

There are two main techniques for growing seeds. One is to sow the seeds directly into the soil and the other is to start them in a punnet before transplanting the seedlings into the soil. Seed packets list useful information including when to sow and the best method to use for the flower.

DIRECT TECHNIQUE

1. Prepare the area you want to plant in by adding organic material.

2. Water the area well prior to sowing.

3. Gently place large seeds evenly spaced on the soil. Sprinkle fine seeds onto the soil's surface as evenly as you can manage.

4. Gently sprinkle a fine layer of soil on top of the seeds; a sieve can be helpful with this. Seeds should be sown as deep as they are long. Fine seeds, such as Poppy, don't need to be covered with soil; they can be sprinkled on top of and 'watered in'.

5. Water the area gently.

WHY DIRECT?

Direct sowing is the best method for plants that have a sensitive root system as there is no need to transplant and disturb them. Also, direct sowing is a great time saver for large areas, as you can broadcast an abundance of seeds rather than carefully transplanting a multitude of seedlings.

. . . .

PUNNET TECHNIQUE

1. Use a clean container or punnet with drainage holes. Fill to the rim with quality seed-raising mix.

2. Evenly place large seeds on top of the seed raising mix or give a light sprinkle of fine seeds.

4. Using a sieve, sprinkle a fine layer of seed raising mix on top of the seeds. Remember, you are aiming to sow the seeds only as deeply as the seeds are long so a fine seed like a Poppy will be fine without a sprinkling on top.

5. Water evenly and gently to avoid dislodging seeds.

WHY PUNNET?

When sowing seedlings into a container you can move them to well-lit and warm positions to assist the germination. You can get a head start on seeds early in the season and give seedlings extra TLC while protecting them from pests in the garden.

PLANTING

SOAK THE ROOTS
Soak the root area in a bucket or wheelbarrow with a dash of seaweed solution but keep it just deep enough to submerge the root system and no higher, for no longer than a few hours. The seaweed acts as a tonic, stimulating the root system and protecting the plant from transplant shock.

TEASE ROOT SYSTEM
Tease the roots of plants apart to free up their ends at the base of the plant. If plant roots have become severely bound or entangled, give them a light prune around the edges.

PLANTING THE PLANT
Dig a hole proportionally larger than the plant you want to plant, incorporating organic matter into the soil. Water the planting hole well. Place the plant and gently backfill with more potting mix or soil to fill the hole. Don't compress around the plant with your hands, as this will remove air pockets. Water well with a diluted seaweed solution.

Many plants are easy to grow from a cutting – especially *Pelargoniums*, *Salvias* and *Sedum* – which means snipping of a section of stem to grow a new plant. Stems with young, green tips and nodes or leaf joints have the greatest potential for forming a new root system. Choose from established plants that look healthy and free of disease, and avoid flowering shoots. Stems with foliage but no flowers are ideal.

The ideal time to take cuttings is from early spring through to early summer, as this is when most plants are naturally going through a growth spurt.

1. Cut the stem below the node (leaf joint). Remove the lower leaves and trim foliage above the node to about a third.
2. Fill clean pots or containers with the cutting mix. Water the pot until there is moisture coming out of the bottom.
3. Make a hole in the cutting mix and gently place the cutting in, nudging the mix around it. Repeat the process until your pot is full of cuttings.
4. Water the foliage with a spray bottle or mister. Place in a warm location – or create your own by placing half a clear plastic bottle or a clear plastic bag over the cuttings.

SEEDLINGS

Water seedlings regularly to ensure they do not dry out, but also check that they are not sitting in water as they are likely to rot if they're too damp. You will get a knack for this balance of watering with practice.

Once the seedlings have formed their first set of leaves, they have usually used up the nutrients and energy enclosed in the seed. They will benefit from a little diluted liquid fertiliser every fortnight until you transplant them into the garden or larger planter.

CUTTINGS

Cuttings need warmth and moisture to produce roots. As the new cutting is without a root system, it must absorb moisture from its remaining foliage. Spray the leaves and stems with a mister or spray bottle two to three times each day and water every couple of days. The cutting mix should be moist but not overly saturated to keep the cutting from rotting. If any cuttings look sick, remove them and any fallen leaves to prevent mould and disease from spreading.

It can take ten days to three weeks for roots to appear on a cutting and then three or four weeks for the root system to be significant enough to pot up the individual plant. Roots poking out of the drainage hole of a pot or container are an excellent sign that it is time to pot the cuttings up into individual pots. A butter knife or stick is a handy tool to carefully lever the new plant up while gently holding on to its leaves. Allow the plants to grow outside in their pots and harden up for a week or two before planting.

TIP PRUNING

Tip pruning, also called pinching out, is a straightforward method that increases the density of stems and flowers on a plant. Tip pruning is most effective while the plant is growing early in the season – ideally before flowering. This technique is also fantastic for making shrubs denser, but it is best to do it when they are young.

By pruning out the growing tip of a plant to the node or leaf joint, it encourages the plant to send out more shoots from the dormant buds near the wound, increasing the volume of flowers. Though it can be done with your fingernails, for a better, cleaner cut, use a sharp pair of secateurs or flower snips with a thin blade. Remove around ½ to ⅓ of the stem above the node (leaf joint) and prune the whole plant to create an even shape.

Flowers that bloom on the tips of their stems, like *Salvias* **(see p. 94)**, *Penstemons* **(see p. 66)**, Everlastings **(see p.106)** and Snapdragons **(see p.46)**, respond well to tip pruning. You can also tip prune flowering climbers like Sweet Peas **(see p. 40)** to increase flowers and foliage.

. . . .

DEADHEADING

Remove flowers which have finished blooming, to encourage new buds and flowers. Choose a node below the finished flower and make a slightly angled cut just above it.

. . . .

CUTTING BACK

After flowering, cut back foliage of a herbaceous perennial to its base as it goes dormant. Cut the dried stems back to the base either during winter or at the early stages of spring when you see new growth starting to emerge.

A decent cut back can encourage new growth after flowering for perennials and shrubs that don't go dormant every year. As a very general rule of thumb, only cut plants back by a third of their full size.

Plants can occasionally get leggy, which might work for your garden, but if you prefer a full look, consider cutting back hard and close to the base to rejuvenate growth during the time of year when the plant is actively growing.

JAC SEMMLER is foremost a plant lover. Known for her big laugh,
Jac tends to Super Bloom – a creative plant practice that brings dynamic
living beauty and diversity to landscapes, places and creative projects,
exploring gardening as an immersive art form. Her first book, *Super Bloom:
A Field Guide to Flowers for Every Gardener*, was published in 2022.

Jac is a respected horticuluralist and recognised innovator in dynamic planting
design. As a plant advocate, Jac shares garden dreams and schemes as a regular
contributor, curator and 'gardener in residence' to institutions, festivals, radio,
podcasts and publications, including *Wonderground Press*.

Jac's Heartland garden is tenderly shared with her partner, Matt, and dog, Melba,
in south-eastern Australia. Jac ponders and cultivates big questions on the
intersections of gardening, art and beauty, testing and contributing to the next
generation of Australian gardens and gardening possibilities.

JAC SEMMLER
@jac.semmler • @superbloomau
hello@thesuperbloom.com.au
www.thesuperbloom.com.au

First published in Australia in 2023
by Thames & Hudson Australia Pty Ltd
11 Central Boulevard, Portside Business Park
Port Melbourne, Victoria 3207
ABN: 72 004 751 964

First published in the United Kingdom in 2024
by Thames & Hudson Ltd
181a High Holborn
London WC1V 7QX

First published in the United States of America
in 2024
by Thames & Hudson Inc.
500 Fifth Avenue
New York, New York 10110

Super Bloom Handbook © Thames & Hudson
Australia 2023

Text © Jac Semmler
Images © Sarah Pannell and Jac Semmler
Illustration © Ashlea O'Neill

26 25 24 23   5 4 3 2 1

Thames & Hudson Australia wishes to acknowledge
that Aboriginal and Torres Strait Islander people are
the first storytellers of this nation and the traditional
custodians of the land on which we live and work.
We acknowledge their continuing culture and pay
respect to Elders past, present and future.

ISBN 978-1-760-76402-9
ISBN 978-1-760-76408-1 (U.S. edition)

British Library Cataloguing-in-Publication Data
A catalogue record for this book is available from
the British Library

Library of Congress Control Number 2023937171

Every effort has been made to trace accurate
ownership of copyrighted text and visual materials
used in this book. Errors or omissions will be
corrected in subsequent editions, provided
notification is sent to the publisher.

Front cover: Ashlea O'Neill I Salt Camp Studio
Design: Ashlea O'Neill I Salt Camp Studio
Editing: Jo Turner
Printed and bound in China by C&C Offset
Printing Co., Ltd

FSC® is dedicated to the promotion of responsible
forest management worldwide. This book is made
of material from FSC®-certified forests and other
controlled sources.

Be the first to know about our new releases,
exclusive content and author events by visiting

thamesandhudson.com.au
thamesandhudson.com
thamesandhudsonusa.com

A catalogue record for this
book is available from the
National Library of Australia